What a day! Today is the
day Roy moves. Will he make
some friends at school?

Roy likes to play with his
cat Tiger. They play together in
the backyard. Tiger runs over
to the soil and digs.

"What can we do today?"
asks Roy.

"Let's go see the stores,"
Mom says.

Mom takes Roy to a toy
store.

Roy looks at many toys in the store. He points to a puppet.

"I like this toy the best," he says. "I can paint this toy!"

Roy's puppet is made from wood. First Roy paints a face on the toy. "I will use oil paint," says Roy.

Mom can fix the puppet's
strings. Then she rubs oil on
the wood.

"What a great toy!" says Roy.

Roy names his toy Super Boy. "Super Boy can point his arm," Roy says.

Now Roy's puppet is a toy that can move!

Roy takes Super Boy to school. His classmates want to play with Super Boy.

The kids in Roy's class enjoy his toy.

The End